MESSENGER P.L. OF NORTH AURORA
3 6878 00113 667

P9-CFU-727

20 Questions: Science

What Do You Know About Animal Adaptations?

PowerKiDS press.
New York

Suzanne Slade

With love to Aunt Jean and Uncle Lee

Published in 2008 by The Rosen Publishing Group, Inc.
29 East 21st Street, New York, NY 10010

First Edition

Editor: Amelie von Zumbusch
Book Design: Kate Laczynski
Photo Researcher: Kate Laczynski

Photo Credits: Cover © Dominic Rouse/Getty Images; p. 5 © www.istockphoto.com/Hazlan Abdul Hakim; pp. 6, 7, 10 (bottom), 12, 13, 14, 16, 19, 21 Shutterstock.com; p. 8 © www.istockphoto.com/Beverly Vycital; p. 9 © www.istockphoto.com/Bruce MacQueen; p. 10 (top) © www.istockphoto.com/Kevin Gryczan; p. 11 © Getty Images; p. 15 (top) © Fred Breummer/Peter Arnold; p. 15 (bottom) © www.istockphoto.com/Geoff Hardy; p. 17 (left) © www.istockphoto.com/Johan Swanepoel; p. 17 (right) © www.istockphoto.com; p. 18 (top) © www.istockphoto.com/Steven Love; p. 18 (bottom) © www.istockphoto.com/Tara Minchin.

Library of Congress Cataloging-in-Publication Data

Slade, Suzanne.
What do you know about animal adaptations? / Suzanne Slade. — 1st ed.
p. cm. — (20 questions : Science)
Includes index.
ISBN 978-1-4042-4199-2 (library binding)
1. Animals—Adaptation—Miscellanea—Juvenile literature. I. Title.

QL49.S55 2008
591.5—dc22

2007033532

Manufactured in the United States of America

Contents

Animal Adaptations

The world is always changing. Animals must also change to live. Nature and people change the places where animals live. Animals adapt, or change, to find food and to stay safe. Some animal adaptations happen quickly, while others take many years.

If an animal changes the way it acts to stay alive, this is called a **behavioral adaptation**. For example, some geese fly south in the fall. This behavior allows them to stay warm and find food during winter. Changes in an animal's body are called **physical adaptations**. A **chameleon** can change color to hide from its enemies. Throughout history, animals have adapted to our changing world.

A giraffe's long neck is one example of a physical adaptation. Its long neck lets a giraffe eat leaves that other animals cannot reach.

1. What are some ways animals change their behavior?

Animals around the world adapt their behavior in different ways to stay safe and comfortable. Animals that live in the cold North often travel to warmer places during winter. This behavioral adaptation is called migration. By migrating south, animals can find food and water to drink. Other northern animals curl up in their beds and sleep through the cold winter. This adaptation is called hibernation.

Sandhill cranes migrate as far as 2,500 miles (4,023 km) each year. These cranes spend their summers in Canada, Alaska, and Siberia and their winters in New Mexico, Texas, and Mexico.

2. What happens when an animal hibernates?

During hibernation, an animal's heart slows and its body **temperature** drops. This means the animal can live on little or no food.

Marmots hibernate each winter. Most marmots hibernate in family groups and share their underground dens with other family members.

3. How do animals get ready for hibernation?

Animals eat lots of food before they hibernate to put extra fat on their body. This fat helps keep them warm.

Black bears eat lots of nuts, berries, fruit, roots, honey, and small animals to get ready to hibernate.

4. Do hibernating animals get hungry?

Some animals wake up during hibernation to eat. They store food, such as berries and nuts, in their homes so they can have a quick bite. Chipmunks are one of the animals that snack during hibernation.

5. How long do animals hibernate?

Every kind of animal hibernates for a different length of time. Painted turtles hibernate for up to four months. Groundhogs sleep in their underground dens for about five months. Bats hibernate in caves for five to six months. Northern black bears can stay in hibernation for over seven months.

Chipmunks hibernate from late fall to early spring. They generally first come out of their dens in March.

6. Do any animals migrate besides birds?

Sea animals such as whales, crabs, eels, **manatees**, and sea turtles are known to migrate. Ants, butterflies, moths, and other bugs migrate, too.

In late spring and early summer, horseshoe crabs migrate from the ocean to North America's eastern shores to lay their eggs.

Pacific green sea turtles migrate hundreds of miles (km) to lay their eggs on warm beaches.

7. How do animals find their way when they migrate?

Animals travel hundreds and sometimes thousands of miles (km) during migration. No one is certain how migrating animals keep from getting lost, but people who study animals have a few ideas. Some believe birds use landmarks such as rivers, mountains, or beaches to find their way. Animals might also be led by certain smells in the air. When looking for direction, animals may use the position of the Sun or stars as a guide. Animals might use one or several of these methods when migrating.

Each year, millions of monarch butterflies migrate thousands of miles (km) to spend the winter in the forests of Michoacán, Mexico.

8. Do animals use adaptations to find food?

Animals use several adaptations to find and catch **prey**. Many bats find bugs to eat by sending out sound waves. The sound waves hit the bugs, return to the bat, and tell it where to find the bugs in the dark. Owls have eyes that can see in the dark, ears that can hear tiny animals, and sharp claws for catching their prey. Frogs use their long, sticky **tongue** to catch their food in one

Some animals use their speed to catch prey. For example, cheetahs can run 70 miles per hour (113 km/h), which is faster than any of the animals they hunt.

A bald eagle's excellent eyesight and sharp claws help it catch its prey. In fact, bald eagles can even spot fish and pull them straight out of the water!

quick flick. Fish called archerfish spit at flying prey. These fish have special adaptations, such as sharp eyes and a powerful tongue, to help them get a tasty dinner almost every time.

9. Can how an animal looks be a useful adaptation?

Some animals are hard to see against their surroundings. This is called **camouflage**. For example, polar bears' white fur makes the bears hard to spot in the snow. This helps them creep up on their prey. Animals also use camouflage to stay safe. Some animals are even camouflaged to look like something else in nature. A stinkbug's flat green shell makes it look like a tiny leaf!

A polar bear's white fur helps it catch its prey. Polar bears mainly hunt seals, but they also eat walruses, whales, caribou, and birds.

10. How else do animals hide?

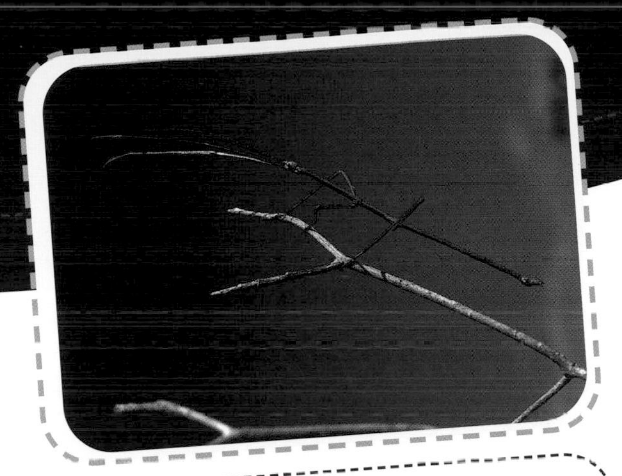

This stick insect, also known as a walking stick, escapes being eaten by looking just like a twig or stem.

Many animals hide by getting out of sight. Animals with sharp claws, such as a mole, can dig holes and hide underground. Turtles carry their hiding place wherever they go. They can suddenly look like a rock.

Hermit crabs keep themselves safe by moving into an empty seashell. As they grow larger, hermit crabs find larger shells to crawl into.

11. What other adaptations do animals use when they are in danger?

Many animals have built-in **weapons**. Some use their claws and teeth to fight off enemies. **Hedgehogs** are covered with long, sharp **spines**. They roll into a ball when danger comes near.

An adult hedgehog generally has between 5,000 and 7,000 spines.

12. Do animals have any weapons their enemies cannot see?

Skunks drive off enemies with a terrible smell. Their stinky weapon floats through the air unseen.

Wildebeests, which live in Africa, escape their enemies by running as fast as 50 miles per hour (80 km/h).

13. Do some animals just run away from their enemies?

Many animals run from their enemies. An animal called the pronghorn escapes most enemies by running as fast as 60 miles per hour (97 km/h)!

14. How do water animals stay out of danger?

Many sea creatures, such as snails, clams, and crabs, have hard shells that keep them safe. **Stingrays**, lionfish, and others have sharp, poison-filled spines.

Lionfish use camouflage and speed to hunt smaller fish. Their poisonous spines are just used to drive off enemies.

15. How do animals in hot places live?

About 5,000 different kinds of animals live in hot, dry deserts. Many desert animals adapt to the heat by sleeping underground during the day. Coyotes, foxes, kangaroo rats, and other animals come out at night when it is cooler.

Black-tailed jackrabbits' huge ears help these animals stay cool. The jackrabbits' blood cools off as it moves through their thin, wide ears.

As jackrabbits do, desert foxes have extra large ears that help them keep cool. These ears also let these foxes hear their prey when they hunt in the cool night.

16. Do desert animals need water?

Kangaroo rats do not drink water at all. These brown and white rats get the water they need by eating seeds. Desert tortoises can last a year without a drink. They get most of their water from the desert flowers and grass they eat. A camel can drink up to 30 gallons (114 l) of water at one time. It will not need water again for several weeks.

Many people think camels store the water they drink in their humps, but these humps really have fat stored in them. Camels actually store extra water in their blood!

17. What special adaptations help animals live in cold places?

Penguins fight the cold with feathers. Their outer feathers keep water out. They also have a second type of feathers called down. Down feathers hold in the heat a penguin's body makes.

Some Arctic animals, such as the walrus, seal, and whale, have extra fat that keeps them warm in icy waters. The bears, wolves, foxes, and rabbits that live in the Arctic have a fur coat, which keeps them warm. Their coat gets thicker in the cold winter months. Arctic foxes and other small animals get out of the cold wind by hiding in holes they dig in the snow.

Penguins, such as these Adélie penguins, have thick blubber, or fat, under their skin. Along with feathers, this blubber keeps the penguins warm.

18. Can living in a group help animals stay alive?

Musk oxen are large, furry animals that live in cold places. They keep warm by staying close together. Several adult lions and their cubs live together in a group, called a pride. When a mother lion catches food, she shares it with the pride. Many fish swim together in schools. It is hard for an enemy to see and catch a fish in a large school.

19. Does every animal adapt to changes?

Most animals in Earth's history have not adapted to changes.

20. What happens to animals that do not adapt?

When animals cannot adapt to changes, they die out, or become extinct. Dinosaurs, dodo birds, and woolly mammoths are just some of the animals that are now extinct.

Glossary

behavioral adaptation (bih-HAY-vyuh-rul a-dap-TAY-shun) A change in an animal's actions that helps it keep living.

camouflage (KA-muh-flahj) Hiding by looking like the things around one.

chameleon (kuh-MEEL-yen) A lizard that can change color to match its surroundings.

hedgehogs (HEJ-hogz) Small animals that have hard spines.

manatees (MA-nuh-teez) Large plant-eating ocean animals.

physical adaptations (FIH-zih-kul a-dap-TAY-shunz) Ways an animal's body has changed that help it keep living.

prey (PRAY) Animals that are hunted by another animal for food.

spines (SPYNZ) Sharp, pointy things.

stingrays (STING-rayz) Flat fish that have spines on their long tail.

temperature (TEM-pur-cher) The heat in a living body.

tongue (TUNG) A part inside the mouth used to eat, make sounds, and swallow.

weapons (WEH-punz) Objects used to hurt, kill, or scare away.

Index

Web Sites

Due to the changing nature of Internet links, PowerKids Press has developed an online list of Web sites related to the subject of this book. This site is updated regularly. Please use this link to access the list:
www.powerkidslinks.com/20sci/adap/